EV CHARGERS

DC FAST CHARGERS' INSTALLATION GUIDE

Dr. Maxwell Shimba

TABLE OF CONTENTS

INTRODUCTION

Understanding DC Fast Chargers

As the demand for electric vehicles (EVs) continues to rise, so does the need for efficient and rapid charging solutions. DC fast chargers, also known as Level 3 chargers, are essential for meeting this demand due to their ability to quickly recharge EV batteries, making them ideal for commercial settings and long-distance travel corridors.

DC fast chargers convert AC power from the grid into DC power, which is then directly supplied to the vehicle's battery. This bypasses the vehicle's onboard charger, allowing for much faster charging times compared to Level 1 and Level 2 chargers. Typically, a DC fast charger can recharge an EV battery to 80% capacity in 20 to 40 minutes, depending on the battery size and the charger's power output.

Key Components of DC Fast Chargers

To understand how DC fast chargers work, it is essential to become familiar with their key components:

1. Rectifier:

- Function: Converts AC power from the grid into DC power.

- Importance: This is a crucial step because EV batteries can only store and utilize DC power.

2. Power Module:

- Function: Regulates the voltage and current supplied to the EV battery.

- Importance: Ensures that the charging process is safe and efficient by providing the correct power levels based on the battery's specifications.

3. Cooling System:

- Function: Keeps the charger and cables at an optimal temperature during high-power charging.

- Importance: Prevents overheating and maintains charger efficiency and safety.

4. Connector and Cable:

- Function: Transfers DC power from the charger to the vehicle.

- Importance: Designed to handle high power levels and ensure a secure and safe connection.

5. User Interface:

- Function: Allows users to interact with the charger, start and stop charging sessions, and monitor charging status.

- Importance: Provides a user-friendly experience and necessary information during the charging process.

Types of DC Fast Chargers

There are several types of DC fast chargers, each with its own specifications and use cases:

1. CHAdeMO:

- Origin: Developed in Japan.

- Compatibility: Used by manufacturers like Nissan and Mitsubishi.

- Power Output: Typically provides up to 50 kW, with newer versions capable of higher outputs.

2. Combined Charging System (CCS):

- Origin: Developed by a consortium of European and American automakers.

- Compatibility: Widely adopted by manufacturers such as BMW, Volkswagen, and General Motors.

- Power Output: Can provide up to 350 kW, making it suitable for ultra-fast charging.

3. Tesla Supercharger:

- Origin: Developed by Tesla Inc.

- Compatibility: Specifically designed for Tesla vehicles, although adapters are available for other EVs.

- Power Output: Can provide up to 250 kW, with plans for higher outputs in the future.

4. GB/T:

- Origin: Developed in China.

- Compatibility: Used primarily in China by local manufacturers.

- Power Output: Offers various power levels, typically up to 180 kW.

Benefits of DC Fast Chargers

DC fast chargers offer numerous benefits that make them essential for the widespread adoption of EVs:

1. Reduced Charging Time:

- Efficiency: Significantly reduces the time required to recharge an EV, making it more convenient for drivers and comparable to refueling a traditional vehicle.

- Travel Convenience: Ideal for long-distance travel, reducing the downtime needed for charging stops.

2. Enhanced User Experience:

- Accessibility: Enables quick and easy access to charging in commercial settings, such as shopping centers, restaurants, and highways.

- Convenience: Improves the overall EV ownership experience by providing more charging options and reducing range anxiety.

3. Support for Fleet Operations:

- Efficiency: Essential for commercial fleets, such as delivery trucks, taxis, and buses, where downtime needs to be minimized.

- Cost Savings: Reduces operational costs by enabling quick turnarounds and maximizing vehicle utilization.

Challenges and Considerations

While DC fast chargers offer many advantages, there are also challenges and considerations to keep in mind:

1. Infrastructure Costs:

- High Installation Costs: The installation of DC fast chargers requires significant investment in electrical infrastructure and site preparation.

- Maintenance Costs: Ongoing maintenance and potential upgrades can add to the overall costs.

2. Grid Impact:

- Demand on the Grid: High-power chargers can place significant demand on the electrical grid, requiring careful planning and coordination with utility companies.

- Energy Management: Advanced energy management systems are needed to ensure stable and efficient grid operation.

3. Compatibility:

- Connector Standards: The presence of multiple charging standards (CHAdeMO, CCS, Tesla, GB/T) requires compatibility solutions and can complicate infrastructure planning.

- Vehicle Support: Not all EVs are equipped to handle the highest power levels offered by DC fast chargers, requiring careful consideration of vehicle capabilities.

Future Trends in DC Fast Charging

The future of DC fast charging is marked by ongoing advancements and innovations:

1. Higher Power Outputs:

- Ultra-Fast Chargers: Development of chargers with power outputs exceeding 350 kW, reducing charging times even further.

- Solid-State Batteries: Advancements in battery technology will support higher power levels and faster charging rates.

2. Wireless Charging:

- Convenience: Wireless DC fast charging systems are being developed, offering seamless and cable-free charging experiences.

- Dynamic Charging: On-the-move charging technologies will enable vehicles to charge while driving, reducing the need for frequent stops.

3. Smart Charging Solutions:

- Grid Integration: Enhanced integration with smart grids will optimize energy use and reduce the impact on the grid.

- V2G Technology: Vehicle-to-grid technology will enable EVs to supply power back to the grid, enhancing grid stability and supporting renewable energy integration.

In conclusion, understanding DC fast chargers is essential for anyone involved in the EV industry, from manufacturers and installers to policymakers and consumers. These chargers play a critical role in the transition to sustainable transportation, offering the speed and efficiency needed to support the growing number of electric vehicles on the road. By staying informed about the latest advancements and best practices, we can ensure the successful deployment and operation of DC fast charging infrastructure, paving the way for a cleaner, greener future.

xii

DR. MAXWELL SHIMBA

UNDERSTANDING DC FAST CHARGERS

As the demand for electric vehicles (EVs) continues to rise, so does the need for efficient and rapid charging solutions. DC fast chargers, also known as Level 3 chargers, are essential for meeting this demand due to their ability to quickly recharge EV batteries, making them ideal for commercial settings and long-distance travel corridors.

DC fast chargers convert AC power from the grid into DC power, which is then directly supplied to the vehicle's battery. This bypasses the vehicle's onboard charger, allowing for much faster charging times compared to Level 1 and Level 2 chargers. Typically, a DC fast charger can recharge an EV battery to 80% capacity in 20 to 40 minutes, depending on the battery size and the charger's power output.

Importance of DC Fast Chargers in EV Infrastructure

The development of a robust EV infrastructure is critical to the widespread adoption of electric vehicles. DC fast chargers play a pivotal role in this infrastructure by providing a quick and convenient charging solution for EV drivers. They are particularly valuable in:

- High-Traffic Areas: DC fast chargers are ideal for locations with high vehicle turnover, such as highway rest stops, shopping centers, and urban centers, where drivers can recharge their vehicles in a short amount of time.

- Fleet Operations: Commercial fleets, such as taxis, delivery vans, and public transportation, benefit greatly from the rapid turnaround times provided by DC fast chargers.

- Range Anxiety Reduction: One of the main concerns for potential EV buyers is range anxiety—the fear of running out of battery power without access to a charger. DC fast chargers help alleviate this concern by offering quick recharging options along major travel routes.

Economic and Environmental Benefits

Investing in DC fast charger installations offers several economic and environmental benefits:

- Revenue Generation: Charging station operators can generate revenue through usage fees, making DC fast chargers a profitable addition to commercial properties.

- Increased Property Value: Installing EV chargers can enhance the attractiveness and value of commercial properties, drawing in EV drivers who may spend time and money at nearby businesses while their vehicles charge.

- Environmental Impact: By supporting the adoption of electric vehicles, DC fast chargers contribute to reducing greenhouse gas emissions and dependence on fossil fuels, promoting a cleaner and more sustainable environment.

In conclusion, DC fast chargers are a crucial component of the EV charging infrastructure, providing rapid and efficient charging solutions that meet the needs of modern EV drivers. This guide will walk you through the entire process of installing a DC fast charger, from site assessment to maintenance, ensuring a successful and smooth installation experience.

CHAPTER 02

SITE ASSESSMENT AND PREPARATION

Selecting the Ideal Location

Choosing the right location for a DC fast charger is crucial for its success and utilization. Several factors should be considered to ensure that the location is both practical and beneficial for EV drivers:

1. Proximity to High-Traffic Areas:

- Identify locations with high vehicular traffic such as shopping centers, highway rest stops, and downtown areas.

- Ensure visibility and ease of access to attract more users.

2. Accessibility:

- The charging station should be easily accessible to all drivers, including those with disabilities.

- Consider the ease of ingress and egress, and ensure there is sufficient space for vehicles to maneuver.

3. Availability of Amenities:

- Choose locations near amenities such as restrooms, restaurants, and retail stores to provide convenience for drivers while they wait for their vehicle to charge.

4. Safety:

- Ensure the area is well-lit and secure to make users feel safe, especially during nighttime.

5. Future Expansion:

- Select a site that has enough space for future expansion if additional charging stations are needed as EV adoption grows.

Permitting and Regulations

Before installing a DC fast charger, it is essential to comply with local, state, and federal regulations. This involves obtaining the necessary permits and adhering to specific guidelines:

1. Zoning and Land Use:

- Verify that the chosen site is zoned for commercial use and that EV charging stations are permitted.

2. Building Permits:

- Apply for building permits from the local government. This may involve submitting detailed site plans and electrical diagrams.

3. Environmental Regulations:

- Ensure that the installation complies with environmental regulations, particularly if the site is near sensitive areas.

4. Utility Company Requirements:

- Coordinate with the local utility company to understand their requirements and ensure that the electrical grid can support the additional load.

5. ADA Compliance:

- Make sure the installation meets the Americans with Disabilities Act (ADA) requirements, ensuring accessibility for all users.

Electrical Requirements

DC fast chargers require a robust electrical infrastructure to function efficiently. Key considerations include:

1. Power Supply:

- Assess the available power supply and ensure it meets the requirements of the DC fast charger. Most DC fast chargers require three-phase power with high voltage and current capacities.

2. Transformer Capacity:

- Check if the existing transformer can handle the additional load or if an upgrade is necessary.

3. Electrical Panel:

- Ensure the electrical panel has enough capacity to accommodate the charger's requirements. This may involve installing a dedicated breaker.

4. Wiring and Conduits:

- Use appropriately rated wiring and conduits to handle the high current levels. This includes considering voltage drop and ensuring proper grounding.

5. Utility Upgrades:

- In some cases, utility upgrades may be required, such as increasing the service size or upgrading the transformer. Work with the utility company to plan and execute these upgrades.

Site Preparation

Proper site preparation is essential to ensure a smooth installation process. This involves:

1. Survey and Layout:

- Conduct a detailed site survey to mark the exact location of the charger and identify any potential obstacles.

- Plan the layout to optimize space utilization and ensure ease of access.

2. Groundwork:

- Prepare the ground by ensuring it is level and stable. This may involve excavation, grading, and pouring a concrete pad.

3. Conduit Installation:

- Install conduits for electrical wiring, ensuring they are properly sealed and protected against environmental factors.

4. Signage and Markings:

- Install appropriate signage to guide users to the charging station. This includes directional signs, usage instructions, and safety warnings.

5. Networking Infrastructure:

- Set up the necessary networking infrastructure to enable remote monitoring and management of the charger. This may involve installing communication cables or setting up wireless connectivity.

In summary, thorough site assessment and preparation are crucial steps in the installation of a DC fast charger. By carefully selecting the location, obtaining the necessary permits, ensuring electrical requirements are met, and preparing the site, you can lay the foundation for a successful and efficient charging station. The next chapter will cover the equipment and tools needed for the installation process.

EQUIPMENT AND TOOLS

Required Equipment for Installation

To successfully install a DC fast charger, you will need a variety of equipment that ensures both functionality and safety. Here are the key components:

1. DC Fast Charger Unit:

- The main unit, which includes the charging cables, connectors, and user interface.

- Ensure compatibility with multiple EV models by including standard connectors such as CCS (Combined Charging System) and CHAdeMO.

2. Electrical Panels and Breakers:

- High-capacity electrical panels and breakers to manage the power supply.

- Use surge protection devices to safeguard against electrical faults.

3. Transformers and Power Distribution Units (PDUs):

- Transformers may be needed to step down or step up the voltage.

- PDUs manage and distribute electrical power to the charger.

4. Wiring and Conduits:

- Heavy-duty wiring capable of handling high current loads.

- Conduits to protect and organize electrical wiring.

5. Grounding and Bonding Materials:

- Proper grounding rods and bonding straps to ensure electrical safety.

6. Networking Equipment:

- Routers, switches, and communication cables for network connectivity.

- Remote monitoring and management systems to oversee charger operations.

7. Mounting Hardware:

- Brackets, bolts, and other mounting accessories to securely install the charger unit.

8. Signage:

- Informational and directional signs to guide users and comply with regulations.

Tools and Safety Gear

Installing a DC fast charger requires specialized tools and safety gear to ensure a safe and efficient process. Here's a list of essential tools and equipment:

1. Hand Tools:

- Screwdrivers, wrenches, pliers, and wire strippers for various tasks.

- Multimeter for measuring voltage, current, and resistance.

2. Power Tools:

- Drills, saws, and grinders for cutting and shaping materials.

- Impact drivers for fastening screws and bolts.

3. Measurement Tools:

- Tape measures and levels for accurate measurements and alignment.

- Laser distance meters for precise site assessment.

4. Electrical Testing Equipment:

- Insulation resistance testers to check the integrity of electrical insulation.

- Circuit testers to verify connections and identify faults.

5. Personal Protective Equipment (PPE):

- Insulated gloves and boots to protect against electrical hazards.

- Safety glasses, hard hats, and high-visibility clothing for general safety.

- Hearing protection for noisy environments.

6. Lifting and Handling Equipment:

- Forklifts, pallet jacks, and cranes for moving heavy equipment.

- Lifting straps and slings for safe handling of large components.

7. Grounding and Bonding Tools:

- Ground resistance testers to ensure proper grounding.

- Bonding clamps and connectors for secure electrical connections.

Safety Considerations

Safety is paramount during the installation of DC fast chargers. Here are some key safety considerations:

1. Electrical Safety:

- Always de-energize circuits before working on them.

- Use lockout/tagout procedures to prevent accidental re-energization.

- Verify the absence of voltage before starting work.

2. Handling Heavy Equipment:

- Use proper lifting techniques and equipment to prevent injuries.

- Ensure that all equipment is securely mounted and stable.

3. Environmental Safety:

- Protect against weather conditions by using weatherproof enclosures and fittings.

- Ensure that all materials and equipment are suitable for outdoor use.

4. Compliance with Regulations:

- Follow all local, state, and federal regulations for electrical installations.

- Ensure that all work is inspected and approved by a qualified electrician or inspector.

In conclusion, having the right equipment and tools is crucial for the successful installation of a DC fast charger. Equally important is adhering to safety protocols to protect both installers and future users. The next chapter will delve into the step-by-step installation process, ensuring you have a clear understanding of each phase.

CHAPTER 04

INSTALLATION PROCESS

Step-by-Step Installation Guide

A successful installation of a DC fast charger requires careful planning and execution. Follow these steps to ensure a smooth process:

1. Site Preparation:

- Clear the Area: Ensure the installation site is free of debris and obstacles.

- Mark the Layout: Use chalk or spray paint to mark the location for the charger, conduits, and electrical panels.

2. Electrical Infrastructure:

- Install Conduits: Lay the conduits for electrical wiring according to the marked layout. Ensure they are properly secured and protected.

- Pull Wires: Run the appropriate wires through the conduits, taking care to use wire pull lubricant if necessary to ease the process.

- Install Transformer (if needed): If a transformer is required, mount it securely and connect it to the main power supply.

3. Mounting the Charger:

- Prepare the Mounting Surface: Ensure the surface is level and stable. Pour a concrete pad if necessary.

- Mount the Charger Unit: Secure the charger unit to the mounting surface using bolts and brackets. Make sure it is stable and aligned.

4. Electrical Connections:

- Connect to Electrical Panel: Attach the wires to the appropriate breakers in the electrical panel. Ensure all connections are tight and secure.

- Grounding: Install grounding rods and connect the charger to the grounding system to ensure safety.

- Bonding: Use bonding straps to connect all metallic parts to prevent electrical shock.

5. Networking Setup:

- Install Communication Equipment: Set up routers, switches, and communication cables to enable network connectivity.

- Configure Remote Monitoring: Ensure the charger is connected to the remote monitoring system for real-time status updates and control.

6. Final Adjustments:

- Check Alignments: Ensure all components are properly aligned and securely mounted.

- Tighten Connections: Double-check all electrical and mechanical connections to ensure they are tight and secure.

Connecting to the Electrical Grid

Connecting the DC fast charger to the electrical grid is a critical step that must be performed with precision:

1. Coordinate with Utility Company:

- Notify Utility Company: Inform the local utility company about the installation and schedule a time for the connection.

- Review Requirements: Ensure all utility company requirements are met, including any necessary upgrades to the electrical service.

2. Install Utility Meter:

- Mount Meter: Install the utility meter in a location specified by the utility company.

- Connect Meter: Connect the meter to the main power supply and the charger's electrical panel.

3. Energize the System:

- Initial Inspection: Have a certified electrician inspect the installation before energizing the system.

- Power On: Once the inspection is complete, energize the system by turning on the breakers and the utility meter.

Safety Measures and Best Practices

Ensuring safety during and after the installation process is paramount. Follow these safety measures and best practices:

1. Personal Protective Equipment (PPE):

- Wear PPE: Always wear appropriate PPE, including insulated gloves, safety glasses, and hard hats.

- Use Fall Protection: If working at heights, use fall protection equipment such as harnesses and lanyards.

2. Electrical Safety:

- De-Energize Circuits: Always de-energize circuits before working on them to prevent electrical shock.

- Verify Absence of Voltage: Use a multimeter to verify the absence of voltage before starting any electrical work.

- Lockout/Tagout Procedures: Follow lockout/tagout procedures to prevent accidental energization.

3. Proper Handling and Lifting:

- Use Lifting Equipment: Use forklifts, pallet jacks, or cranes to move heavy equipment. Do not attempt to lift heavy objects manually.

- Follow Safe Lifting Techniques: When lifting manually, use proper techniques to prevent injury. Bend your knees and keep your back straight.

4. Environmental Considerations:

- Weatherproof Equipment: Ensure all outdoor equipment is weatherproof and protected against the elements.

- Secure Cables: Use cable ties and clamps to secure cables and prevent tripping hazards.

5. Compliance and Inspection:

- Follow Regulations: Adhere to all local, state, and federal regulations regarding electrical installations.

- Inspection and Testing: Have the installation inspected and tested by a qualified electrician before use.

In conclusion, following a detailed step-by-step installation guide and adhering to safety measures are crucial for the successful deployment of a DC fast charger. The next chapter will cover the testing and commissioning process, ensuring that the charger is ready for use and operates safely and efficiently.

CHAPTER 05

TESTING AND COMMISSIONING

Initial Testing Procedures

Once the DC fast charger is installed, it is essential to conduct thorough testing to ensure it operates correctly and safely. Follow these initial testing procedures:

1. Visual Inspection:

- Check Connections: Inspect all electrical connections to ensure they are secure and properly tightened.

- Inspect Cables: Verify that all cables are properly routed and protected against physical damage.

- Verify Grounding: Ensure that all grounding and bonding connections are correctly installed.

2. Pre-Power Tests:

- Continuity Testing: Use a multimeter to test the continuity of all wiring, ensuring there are no breaks or short circuits.

- Insulation Resistance: Perform insulation resistance testing to verify the integrity of the electrical insulation.

3. Power-Up Procedures:

- Initial Energization: Slowly energize the system by turning on the main breaker. Observe for any signs of issues such as unusual noises or smells.

- Voltage Verification: Use a multimeter to check the voltage levels at various points in the system, ensuring they match the expected values.

4. Functionality Tests:

- Charger Display: Verify that the charger's display and user interface are functioning correctly.

- Connectivity: Check the network connectivity to ensure the charger can communicate with remote monitoring systems.

Troubleshooting Common Issues

During the testing phase, you may encounter some common issues. Here are troubleshooting steps for some of the most frequent problems:

1. No Power to Charger:

- Check Breakers: Ensure all breakers are in the "on" position.

- Inspect Wiring: Verify that all wiring connections are secure and properly terminated.

- Check Utility Supply: Confirm that the utility power supply is active and within the correct voltage range.

2. Display Issues:

- Reboot Charger: Try rebooting the charger by cycling the power.

- Check Connections: Verify that all display cables are securely connected.

- Update Firmware: Ensure the charger firmware is up to date.

3. Network Connectivity Problems:

- Check Network Cables: Inspect all network cables for proper connection and potential damage.

- Restart Network Equipment: Restart routers and switches to reset network connections.

- Configure Settings: Verify that the charger's network settings are correctly configured.

4. Charging Errors:

- Inspect Connector: Check the charging connector for damage or debris.

- Test with Different EV: Try charging a different EV to determine if the issue is with the charger or the vehicle.

- Consult Manual: Refer to the charger's manual for specific error codes and troubleshooting steps.

Final Commissioning Steps

After resolving any issues and ensuring the system is functioning correctly, complete the following commissioning steps:

1. Load Testing:

- Simulate Load: Connect an EV to the charger and monitor the charging process to ensure it handles the load correctly.

- Observe Performance: Check the charger's performance metrics such as charging speed and power output.

2. System Calibration:

- Adjust Settings: Calibrate any necessary settings on the charger, such as output voltage and current limits.

- Verify Accuracy: Ensure that the charger's output matches the specified parameters.

3. User Training:

- Demonstrate Operation: Provide a demonstration to potential users on how to operate the charger, including starting and stopping a charging session.

- Safety Instructions: Explain safety procedures and emergency shutdown instructions.

4. Documentation:

- Record Settings: Document all final settings and calibration data for future reference.

- Complete Forms: Fill out any required commissioning forms and submit them to the relevant authorities.

5. Final Inspection:

- Third-Party Inspection: If required, arrange for a third-party inspection to verify that the installation meets all regulatory and safety standards.

- Issue Certification: Obtain any necessary certifications or approvals to confirm the charger is ready for public use.

In conclusion, thorough testing and commissioning are essential to ensure that the DC fast charger operates safely and efficiently. By following these procedures, you can identify and resolve any issues before the charger is put into service, ensuring a smooth and reliable operation. The next chapter will cover maintenance and upgrades to keep the charger in optimal condition.

CHAPTER 06

MAINTENANCE AND UPGRADES

Routine Maintenance

To ensure the longevity and reliability of DC fast chargers, regular maintenance is essential. Here are key routine maintenance tasks:

1. Visual Inspections:

- Check for Damage: Inspect the charger unit, cables, and connectors for any signs of physical damage.

- Inspect Grounding: Verify that grounding connections remain intact and corrosion-free.

- Cleanliness: Ensure the charging station and surrounding area are clean and free of debris.

2. Electrical Checks:

- Test Voltage Levels: Regularly measure voltage levels to ensure they remain within the specified range.

- Inspect Connections: Check all electrical connections for signs of wear, looseness, or corrosion.

- Check Insulation: Perform insulation resistance tests to ensure electrical insulation is intact.

3. Functional Testing:

- Verify Operation: Periodically test the charger's functionality by simulating a charging session.

- Check Display and Controls: Ensure the display and user controls are functioning correctly.

- Test Network Connectivity: Verify that the charger remains connected to the network and can communicate with remote monitoring systems.

4. Firmware and Software Updates:

- Update Firmware: Regularly check for and install firmware updates provided by the manufacturer to ensure the charger operates with the latest features and fixes.

- Review Software: Update any associated software or management systems to maintain compatibility and security.

Preventive Maintenance

Preventive maintenance helps avoid potential problems before they occur. Here are some preventive measures:

1. Scheduled Inspections:

- Regular Checks: Schedule comprehensive inspections at regular intervals (e.g., monthly, quarterly) to identify and address potential issues.

- Professional Assessment: Consider having a professional technician conduct detailed inspections and maintenance.

2. Component Replacements:

- Replace Worn Parts: Replace components such as connectors, cables, and protective covers that show signs of wear and tear.

- Update Hardware: Upgrade hardware components like circuit breakers and surge protectors as needed to maintain optimal performance.

3. Environmental Protection:

- Weatherproofing: Ensure that all weatherproofing measures remain effective, especially after severe weather conditions.

- Pest Control: Take steps to prevent pests from damaging cables and equipment.

4. Performance Monitoring:

- Remote Monitoring: Use remote monitoring systems to continuously track the performance and status of the charger.

- Analyze Data: Regularly review performance data to identify trends and potential issues before they escalate.

Upgrades and Enhancements

As technology advances, upgrading and enhancing DC fast chargers can improve their performance and extend their lifespan. Here are some key upgrades and enhancements:

1. Hardware Upgrades:

- Higher Power Output: Upgrade to chargers with higher power output capabilities to reduce charging times.

- Advanced Connectors: Replace older connectors with newer, more efficient models to support a wider range of EVs.

2. Software Enhancements:

- Smart Charging Features: Implement software upgrades that enable smart charging features such as load balancing and energy management.

- User Interface Improvements: Enhance the user interface for better usability and accessibility.

3. Networking Upgrades:

- Faster Connectivity: Upgrade to faster and more reliable network connections to improve remote monitoring and management.

- Enhanced Security: Implement advanced security protocols to protect the charger from cyber threats.

4. Environmental and Sustainability Improvements:

- Solar Integration: Integrate solar panels to power the charger with renewable energy, reducing operational costs and environmental impact.

- Battery Storage: Add battery storage systems to store excess energy and provide backup power during outages.

Documentation and Record-Keeping

Maintaining detailed records of all maintenance and upgrades is crucial for accountability and future reference:

1. Maintenance Logs:

- Record Dates: Document the dates of all inspections, maintenance tasks, and component replacements.

- Detail Work Done: Include detailed descriptions of the work performed and any issues identified.

2. Upgrade Records:

- Document Upgrades: Record all hardware and software upgrades, including dates, descriptions, and any associated costs.

- Track Performance: Monitor and document the performance improvements resulting from upgrades.

3. Compliance Records:

- Regulatory Compliance: Keep records of all inspections and certifications to ensure compliance with local, state, and federal regulations.

- Safety Documentation: Maintain records of all safety checks and incidents to enhance future safety protocols.

In conclusion, regular maintenance and timely upgrades are essential to keep DC fast chargers operating efficiently and safely. By following these guidelines, you can ensure the reliability and longevity of your charging infrastructure, providing a valuable service to EV users. The next chapter will discuss customer support and managing user experience to ensure a positive interaction with your charging stations.

CHAPTER 07

CUSTOMER SUPPORT AND USE EXPERIENCE

Enhancing User Experience

Providing a positive user experience is crucial for the success of DC fast chargers. Here are strategies to enhance user satisfaction:

1. User-Friendly Design:

- Clear Instructions: Ensure the charger's user interface provides clear, easy-to-follow instructions.

- Intuitive Interface: Design the interface to be intuitive, minimizing the learning curve for new users.

- Accessibility: Ensure the charger is accessible to all users, including those with disabilities.

2. Efficient Charging Process:

- Fast Charging: Optimize the charging process to minimize waiting times.

- Reliability: Ensure the charger is reliable and minimizes downtime through regular maintenance.

3. Supportive Features:

- Real-Time Updates: Provide real-time updates on the charging status via the charger's display or a mobile app.

- Payment Options: Offer multiple payment options, including contactless payment methods.

- Customer Feedback: Implement a system for collecting and addressing customer feedback.

Customer Support Services

Effective customer support is essential for resolving issues and maintaining user trust. Here are key components of customer support services:

1. Help Desk:

- 24/7 Support: Offer round-the-clock customer support through phone, email, and chat.

- Knowledgeable Staff: Ensure support staff are well-trained and knowledgeable about the charger's operation and common issues.

2. Online Resources:

- FAQs and Tutorials: Provide a comprehensive FAQ section and step-by-step tutorials on your website.

- User Manuals: Make user manuals and troubleshooting guides easily accessible online.

3. Mobile App Support:

- App Functionality: Develop a mobile app that allows users to locate chargers, check availability, start/stop charging sessions, and make payments.

- In-App Support: Include an in-app support feature for reporting issues and getting help.

4. Remote Diagnostics:

- Proactive Monitoring: Use remote monitoring tools to proactively identify and address issues before they affect users.

- Remote Assistance: Provide remote assistance to troubleshoot and resolve problems without requiring on-site visits.

Handling Common User Issues

Addressing common user issues promptly and effectively is crucial for maintaining user satisfaction. Here are some common issues and their solutions:

1. Charger Unavailability:

- Real-Time Updates: Ensure the mobile app and website provide real-time updates on charger availability.

- Queue Management: Implement a queue management system to manage high demand times.

2. Payment Problems:

- Multiple Payment Methods: Offer various payment methods to accommodate different user preferences.

- Support for Payment Issues: Provide quick support for resolving payment-related issues, such as declined transactions or refund requests.

3. Charging Errors:

- Clear Error Messages: Ensure that error messages are clear and provide guidance on how to resolve the issue.

- Support Contact: Include a support contact number or chat option directly on the charger interface for immediate assistance.

4. Slow Charging Speeds:

- Communicate Limitations: Inform users of any factors that might affect charging speeds, such as high usage or temporary power limitations.

- Maintenance Checks: Regularly inspect and maintain chargers to ensure they operate at optimal speeds.

Building a Community

Creating a community around your charging stations can foster loyalty and encourage positive user interactions:

1. User Forums:

- Online Community: Create online forums or social media groups where users can share experiences, tips, and feedback.

- Engagement: Actively engage with the community to address concerns and provide updates.

2. Events and Promotions:

- User Events: Host events such as launch parties, informational sessions, or user meetups.

- Promotions: Offer promotions or discounts to loyal users or during special occasions.

3. Feedback Loops:

- Regular Surveys: Conduct regular user surveys to gather feedback and identify areas for improvement.

- Implement Changes: Show users that their feedback is valued by implementing changes based on their suggestions.

Training and Education

Educating users on how to properly use DC fast chargers can enhance their experience and reduce issues:

1. User Training Sessions:

- Workshops: Offer workshops or training sessions to educate users on how to use the chargers effectively.

- Virtual Training: Provide virtual training sessions or webinars for users who cannot attend in person.

2. Educational Materials:

- Instructional Videos: Create instructional videos that demonstrate how to use the chargers and troubleshoot common issues.

- Printed Materials: Distribute brochures or quick-start guides at the charging stations.

In conclusion, prioritizing customer support and enhancing the user experience are key to the success of DC fast chargers. By providing excellent support services, addressing common issues, building a community, and offering training and education, you can ensure a positive experience for all users. The next chapter will focus on regulatory compliance and ensuring your charging stations meet all necessary standards and requirements.

CHAPTER 08

REGULATORY COMPLIANCE AND STANDARDS

Understanding Regulatory Requirements

Compliance with local, state, and federal regulations is essential for the operation of DC fast chargers. Understanding and adhering to these regulations ensures safety, reliability, and legal operation. Key areas to consider include:

1. Electrical Codes:

- National Electrical Code (NEC): Ensure installations comply with the NEC, which sets standards for safe electrical design, installation, and inspection.

- Local Electrical Codes: Adhere to local electrical codes that may have additional or more stringent requirements.

2. Building Codes:

- Accessibility Requirements: Ensure compliance with the Americans with Disabilities Act (ADA) to provide accessible charging stations for all users.

- Structural Standards: Meet building codes related to the structural integrity of the charging station installation.

3. Zoning and Permitting:

- Zoning Laws: Verify that the installation site complies with local zoning laws, which dictate land use and may include specific requirements for EV chargers.

- Permits: Obtain all necessary permits before beginning installation, including electrical, building, and environmental permits.

4. Environmental Regulations:

- Environmental Impact: Assess and mitigate any environmental impact, such as potential soil or water contamination.

- Sustainability Practices: Follow regulations promoting sustainability, such as integrating renewable energy sources or using environmentally friendly materials.

Safety Standards

Adhering to safety standards is crucial for preventing accidents and ensuring user safety. Key safety standards include:

1. UL Certification:

- UL Listed Equipment: Use chargers and components that are UL (Underwriters Laboratories) listed, ensuring they meet rigorous safety standards.

- Regular Testing: Periodically test and inspect UL certified equipment to ensure ongoing compliance.

2. Fire Safety:

- Fire Suppression Systems: Install fire suppression systems where necessary to prevent and control electrical fires.

- Emergency Shutdown: Ensure chargers have emergency shutdown features to quickly power off in case of a fire or other emergency.

3. Electrical Safety:

- Ground Fault Protection: Implement ground fault protection to prevent electrical shock and equipment damage.

- Surge Protection: Install surge protection devices to safeguard against voltage spikes.

Certification and Approval Processes

Obtaining the ecessary certifications and approvals is essential for legal operation. Follow these steps:

1. Application Process:

- Submit Documentation: Provide detailed documentation, including site plans, electrical schematics, and equipment specifications, to regulatory authorities.

- Pay Fees: Pay any required application or inspection fees as part of the certification process.

2. Inspections:

- Pre-Installation Inspection: Arrange for a pre-installation inspection to ensure the site and plans meet regulatory standards.

- Post-Installation Inspection: Schedule a post-installation inspection to verify that the installation complies with all applicable codes and standards.

3. Certification:

- Obtain Certificates: Receive certification or approval from regulatory authorities, confirming that the installation meets all required standards.

- Display Certificates: Display certification documents at the charging station or keep them readily available for inspection.

Compliance Monitoring and Reporting

Ongoing compliance monitoring and reporting are essential for maintaining certification and ensuring continuous adherence to regulations:

1. Regular Audits:

- Internal Audits: Conduct regular internal audits to review compliance with all relevant codes and standards.

- External Audits: Arrange for periodic external audits by qualified third parties to verify ongoing compliance.

2. Record Keeping:

- Maintain Records: Keep detailed records of all inspections, maintenance, upgrades, and compliance activities.

- Documentation: Ensure all documentation is up-to-date and readily available for regulatory review.

3. Incident Reporting:

- Report Incidents: Promptly report any incidents, such as electrical faults or accidents, to the relevant authorities.

- Corrective Actions: Implement corrective actions to address any identified issues and prevent recurrence.

Staying Up-to-Date with Regulations

Regulations and standards can change over time. Staying informed and adapting to new requirements is crucial:

1. Industry Associations:

- Join Associations: Become a member of industry associations, such as the Electric Vehicle Charging Association (EVCA), to stay informed about regulatory changes.

- Participate in Forums: Engage in industry forums and discussions to share knowledge and learn from peers.

2. Continuing Education:

- Attend Workshops: Participate in workshops, webinars, and training sessions on regulatory compliance and best practices.

- Certification Renewal: Ensure certifications and licenses are renewed as required, and stay updated on new certification requirements.

3. Regulatory Updates:

- Monitor Updates: Regularly monitor regulatory updates from government agencies and industry bodies.

- Implement Changes: Promptly implement any necessary changes to ensure ongoing compliance with new regulations.

In conclusion, regulatory compliance and adherence to standards are fundamental to the safe and legal operation of DC fast chargers. By understanding and following the necessary requirements, obtaining the required certifications, and maintaining rigorous compliance monitoring, you can ensure your charging stations meet all legal and safety standards. The next chapter will explore the financial aspects of DC fast charger installations, including funding, incentives, and return on investment.

CHAPTER 09

FUTURE TRENDS IN EV CHARGING

9.1 Advancements in Charging Technology

As electric vehicles (EVs) become more prevalent, advancements in charging technology are crucial to meeting the growing demand for efficient, convenient, and reliable charging solutions. The future of EV charging promises significant improvements, particularly in faster charging solutions and wireless charging technology.

Faster Charging Solutions

The need for faster charging solutions is driven by the desire to reduce charging times, making EVs more convenient and comparable to traditional refueling experiences. Here are some key advancements:

1. Ultra-Fast Chargers:

- Higher Power Output: Ultra-fast chargers with power outputs exceeding 350 kW are being developed, capable of adding hundreds of miles of range in just a few minutes. These chargers use advanced cooling systems and high-capacity cables to safely deliver high power levels.

- Wide Adoption: As battery technology improves, more EV models will support ultra-fast charging, making it a standard feature rather than a premium option.

2. Battery Technology Improvements:

- Solid-State Batteries: The development of solid-state batteries promises higher energy densities, faster charging times, and improved safety. These batteries can handle higher charging currents without degrading as quickly as traditional lithium-ion batteries.

- Thermal Management: Advanced thermal management systems are being integrated into EVs to maintain optimal battery temperatures during fast charging, enhancing efficiency and longevity.

3. Modular Charging Stations:

- Scalable Solutions: Modular charging stations allow operators to add more charging modules as demand increases, providing a scalable and cost-effective solution for expanding charging infrastructure.

- Interoperability: These stations are designed to be compatible with various EV models and charging standards, ensuring broad usability and convenience for users.

Wireless Charging

Wireless charging technology is set to revolutionize the way EVs are charged, offering unparalleled convenience and efficiency. Key developments in this area include:

1. Inductive Charging:

- Embedded Infrastructure: Inductive charging systems use electromagnetic fields to transfer energy between a charging pad on the ground and a receiver on the vehicle. These systems can be embedded in parking spaces, driveways, and even roadways.

- Seamless Experience: Users can simply park their EVs over the charging pad to initiate charging, eliminating the need for cables and connectors.

2. Dynamic Charging:

- On-the-Move Charging: Dynamic wireless charging systems enable EVs to charge while driving over specially equipped roads. This technology extends the range of EVs and reduces the need for frequent stops, particularly beneficial for long-haul and public transit vehicles.

- Energy Efficiency: Advanced power management systems ensure efficient energy transfer, minimizing losses and maximizing charging speed.

3. Standardization and Safety:

- Global Standards: Industry stakeholders are working on establishing global standards for wireless charging to ensure compatibility and safety across different manufacturers and regions.

- Health and Safety: Rigorous testing and certification processes are in place to ensure that wireless charging systems are safe for users and do not interfere with other electronic devices.

9.2 Impact of EV Charging on the Grid

The widespread adoption of EVs presents both challenges and opportunities for the electrical grid. Effective integration of EV charging infrastructure with the grid is essential to ensure stability, efficiency, and sustainability.

Smart Grid Integration

Smart grid technology plays a crucial role in managing the increased demand for electricity due to EV charging. Key aspects of smart grid integration include:

1. Grid Modernization:

- Advanced Metering Infrastructure (AMI): Smart meters provide real-time data on electricity consumption, enabling utilities to monitor and manage energy use more effectively. This data is crucial for predicting and balancing the load from EV charging.

- Automation and Control: Automated control systems can dynamically adjust the power supply to charging stations based on real-time demand and grid conditions, optimizing energy distribution and minimizing strain on the grid.

2. Vehicle-to-Grid (V2G) Technology:

- Bidirectional Charging: V2G technology allows EVs to not only draw power from the grid but also supply power back to it. This bidirectional flow can be used to balance grid load during peak demand periods, acting as a distributed energy resource.

- Energy Storage: EVs equipped with V2G capabilities can serve as mobile energy storage units, helping to stabilize the grid and integrate renewable energy sources by storing excess energy during low demand and supplying it during high demand.

3. Renewable Energy Integration:

- Solar and Wind Power: Integrating renewable energy sources with EV charging infrastructure helps reduce the carbon footprint of EVs and promotes the use of clean energy. Smart grid systems can manage the variability of renewable energy generation and ensure efficient use.

- Microgrids: Localized microgrids with renewable energy generation and storage capabilities can support EV charging infrastructure, providing resilience and reducing dependence on the central grid.

Demand Response Programs

Demand response programs are essential for managing the impact of EV charging on the grid, particularly during peak demand periods. These programs incentivize users to adjust their charging habits to support grid stability. Key components include:

1. Time-of-Use Pricing:

- Variable Rates: Utilities offer variable electricity rates based on the time of day, encouraging users to charge their EVs during off-peak hours when electricity is cheaper and demand is lower.

- Load Shifting: By shifting charging to off-peak times, users can reduce their energy costs and alleviate pressure on the grid during peak periods.

2. Incentive Programs:

- Rebates and Discounts: Utilities provide rebates, discounts, or other financial incentives for users who participate in demand response programs and adhere to recommended charging schedules.

- Smart Charging Incentives: Programs that offer incentives for the use of smart chargers, which can automatically adjust charging times and power levels based on grid conditions.

3. Real-Time Demand Management:

- Dynamic Load Control: Smart grid systems can dynamically control the charging power of EVs in real-time, reducing the load during peak demand and increasing it during low demand periods.

- User Notifications: Users receive real-time notifications and recommendations on the optimal times to charge their EVs, helping them take advantage of lower rates and support grid stability.

In conclusion, the future of EV charging is marked by rapid advancements in technology and significant impacts on the electrical grid. Faster charging solutions and wireless charging will enhance the convenience and efficiency of EV charging, while smart grid integration and demand response programs will ensure the grid can support the growing number of EVs sustainably and reliably. Embracing these trends will be crucial for the continued growth and success of the EV ecosystem.

CHAPTER 10

HIGHER POWER OUTPUT

Introduction to Higher Power Outputs

As the electric vehicle (EV) market continues to expand, the demand for faster and more efficient charging solutions is growing. Higher power outputs in EV chargers represent a significant advancement in charging technology, allowing for reduced charging times and greater convenience for EV owners. Two key components driving this progress are the development of ultra-fast chargers with power outputs exceeding 350 kW and the advancements in solid-state batteries that support higher power levels and faster charging rates. This chapter delves into the details of these innovations, their benefits, and the challenges they pose.

Ultra-Fast Chargers: Reducing Charging Times

Ultra-fast chargers are at the forefront of EV charging technology, offering significantly higher power outputs compared to traditional chargers. These chargers are designed to meet the needs of modern EVs and the increasing expectations of consumers for rapid charging solutions.

1. Characteristics of Ultra-Fast Chargers:

- Power Output: Ultra-fast chargers typically offer power outputs exceeding 350 kW, compared to 50-150 kW for standard DC fast chargers. This high power output allows for much faster charging times.

- Cooling Systems: To handle the high power levels, ultra-fast chargers are equipped with advanced cooling systems. These systems prevent overheating and ensure the safe and efficient operation of the charger and the vehicle's battery.

- Connector Types: Ultra-fast chargers use advanced connector types, such as CCS (Combined Charging System) and CHAdeMO, which are capable of handling higher power levels and ensuring compatibility with a wide range of EVs.

2. Benefits of Ultra-Fast Charging:

- Reduced Charging Times: Ultra-fast chargers can significantly reduce charging times, allowing EVs to gain a substantial amount of charge in a matter of minutes. For example, an EV can be charged to 80% capacity in 10-20 minutes, depending on the battery size and initial charge level.

- Convenience for Long-Distance Travel: The reduced charging times offered by ultra-fast chargers make them ideal for long-distance travel, minimizing the downtime for EV owners and making road trips more feasible and convenient.

- Increased Adoption Rates: Faster charging times can alleviate range anxiety and make EVs more attractive to consumers, potentially accelerating the adoption rates of electric vehicles.

3. Implementation of Ultra-Fast Chargers:

- Infrastructure Development: The deployment of ultra-fast chargers requires significant investment in infrastructure, including the installation of high-capacity power lines and transformers. Collaboration between governments, utilities, and private companies is essential to develop and expand the ultra-fast charging network.

- Site Selection: Strategic placement of ultra-fast chargers along major highways and in urban centers is crucial to maximize their utility and accessibility for EV owners.

- Integration with Renewable Energy: Integrating ultra-fast chargers with renewable energy sources, such as solar and wind power, can enhance sustainability and reduce the environmental impact of EV charging.

Solid-State Batteries: Supporting Higher Power Levels

The development of solid-state batteries represents a significant leap forward in battery technology, offering numerous advantages over traditional lithium-ion batteries, particularly in terms of supporting higher power levels and faster charging rates.

1. Characteristics of Solid-State Batteries:

- Solid Electrolytes: Unlike conventional lithium-ion batteries, which use liquid electrolytes, solid-state batteries use solid electrolytes. This change improves safety by reducing the risk of leaks and fires.

- Higher Energy Density: Solid-state batteries offer higher energy density, meaning they can store more energy in a smaller and lighter package. This characteristic is particularly beneficial for EVs, as it can increase driving range without adding significant weight.

- Enhanced Stability and Longevity: Solid-state batteries exhibit greater stability and longer lifespan compared to traditional batteries, which can degrade over time and with repeated charging cycles.

2. Benefits of Solid-State Batteries:

- Faster Charging Rates: The improved properties of solid-state batteries enable them to support much higher charging rates, reducing the time needed to recharge an EV significantly.

- Increased Driving Range: The higher energy density of solid-state batteries translates to longer driving ranges for EVs, addressing one of the major concerns of potential EV buyers.

- Enhanced Safety: Solid-state batteries are less prone to thermal runaway and other safety issues associated with liquid electrolytes, making them a safer option for EVs.

3. Challenges and Considerations:

- Manufacturing Complexity: The production of solid-state batteries involves complex manufacturing processes that are currently more expensive than those for traditional lithium-ion batteries. Scaling up production and reducing costs are critical challenges to address.

- Material Availability: The availability of suitable materials for solid-state batteries, such as lithium and solid electrolytes, can impact their production and widespread adoption.

- Compatibility with Existing Infrastructure: Ensuring that solid-state batteries are compatible with existing charging infrastructure and EV designs is essential for a smooth transition and adoption.

4. Future Prospects and Innovations:

- Research and Development: Ongoing research and development efforts aim to improve the performance, cost-effectiveness, and scalability of solid-state batteries. Advances in materials science and battery engineering are key to unlocking the full potential of this technology.

- Collaborative Efforts: Collaboration between battery manufacturers, automotive companies, and research institutions can accelerate the development and commercialization of solid-state batteries, bringing their benefits to market more quickly.

- Regulatory Support: Government policies and incentives can support the development and adoption of solid-state batteries by funding research, providing subsidies, and setting standards for safety and performance.

Conclusion

The advancements in higher power outputs for EV chargers, including the development of ultra-fast chargers and solid-state batteries, are set to revolutionize the EV charging landscape. These innovations offer the promise of significantly reduced charging times, greater convenience for EV owners, and enhanced driving range, all of which are crucial for the continued growth and acceptance of electric vehicles.

By addressing the challenges associated with infrastructure development, manufacturing complexity, and material availability, stakeholders can ensure that these technologies are effectively integrated into the existing EV ecosystem. The future of EV charging lies in the continued pursuit of higher power outputs, supported by cutting-edge battery technology and strategic infrastructure investments, paving the way for a more efficient, sustainable, and user-friendly electric vehicle experience.

CHAPTER 11

WIRELESS CHARGING

Introduction to Wireless Charging

As the electric vehicle (EV) market continues to evolve, innovations in charging technology are playing a crucial role in shaping the future of sustainable transportation. Among these advancements, wireless charging stands out for its potential to provide unparalleled convenience and efficiency. Wireless DC fast charging systems offer a seamless, cable-free charging experience, while dynamic charging technologies enable vehicles to recharge while on the move, significantly reducing the need for frequent stops. This chapter explores the concepts, benefits, and challenges associated with wireless charging for EVs.

Convenience: Seamless and Cable-Free Charging

Wireless charging, also known as inductive charging, eliminates the need for physical connectors and cables, providing a more convenient and user-friendly charging experience. Here's how it works:

1. Basic Principles of Inductive Charging:

- Electromagnetic Induction: Wireless charging relies on electromagnetic induction to transfer energy between a charging pad (transmitter) and a receiver installed in the vehicle. When the charging pad generates an alternating magnetic field, it induces an electric current in the receiver, which is then used to charge the EV's battery.

- Resonant Inductive Coupling: For efficient energy transfer, both the transmitter and receiver are tuned to the same resonant frequency. This maximizes the amount of energy transferred and minimizes losses.

2. Components of a Wireless Charging System:

- Charging Pad: Installed on the ground or embedded in the pavement, the charging pad generates the magnetic field required for wireless charging.

- Vehicle Receiver: Integrated into the underside of the vehicle, the receiver captures the magnetic field and converts it into electrical energy to charge the battery.

- Control Unit: Manages the charging process, ensuring safe and efficient energy transfer. It communicates with the vehicle's onboard systems to regulate the charging rate and monitor battery status.

3. Advantages of Wireless Charging:

- Ease of Use: Wireless charging eliminates the need for plugging and unplugging cables, making the charging process more convenient, especially in adverse weather conditions.

- Reduced Wear and Tear: By eliminating physical connectors, wireless charging reduces wear and tear on charging equipment, potentially lowering maintenance costs.

- Enhanced Safety: Wireless systems are designed to automatically shut off if foreign objects are detected on the charging pad, enhancing safety.

Dynamic Charging: Charging on the Move

Dynamic wireless charging takes the concept of inductive charging a step further by enabling vehicles to charge while in motion. This innovative approach has the potential to revolutionize the way we think about EV charging:

1. How Dynamic Charging Works:

- Embedded Charging Infrastructure: Dynamic charging systems require charging infrastructure to be embedded in roadways. These systems create a continuous magnetic field that can transfer energy to vehicles as they pass over them.

- Vehicle Adaptations: Vehicles must be equipped with dynamic receivers capable of capturing energy from the roadway and converting it into usable electrical power.

2. Benefits of Dynamic Charging:

- Extended Driving Range: Dynamic charging can significantly extend the driving range of EVs by providing continuous energy replenishment, reducing the need for frequent stops at charging stations.

- Smaller Batteries: With the ability to charge on the go, EVs can be designed with smaller, lighter batteries, reducing vehicle weight and cost.

- Reduced Charging Infrastructure: By incorporating charging capabilities into existing roadways, dynamic charging can reduce the need for extensive charging station networks, simplifying infrastructure development.

3. Applications of Dynamic Charging:

- Public Transit Systems: Dynamic charging is particularly well-suited for public transit systems, such as buses and trams, which follow fixed routes and can benefit from continuous energy replenishment.

- Highways and Major Roads: Implementing dynamic charging on highways and major roads can support long-distance travel and reduce range anxiety for EV owners.

Challenges and Considerations

While wireless and dynamic charging offer exciting possibilities, several challenges and considerations must be addressed to achieve widespread adoption:

1. Infrastructure Development:

- Cost: Developing and deploying wireless and dynamic charging infrastructure can be costly, requiring significant investment from both public and private sectors.

- Standardization: Establishing industry standards for wireless charging is crucial to ensure compatibility between different vehicles and charging systems.

2. Efficiency and Energy Loss:

- Energy Transfer Efficiency: While wireless charging technology has improved, it is generally less efficient than wired charging, resulting in some energy loss during transfer. Ongoing research aims to minimize these losses and improve overall system efficiency.

- Power Levels: Achieving high power levels necessary for fast charging wirelessly can be challenging and requires advanced materials and technologies.

3. Safety and Environmental Concerns:

- Electromagnetic Fields: Ensuring that wireless charging systems do not pose health risks due to electromagnetic fields is a critical consideration. Rigorous testing and regulatory compliance are essential to address these concerns.

- Environmental Impact: Installing dynamic charging infrastructure in roadways may have environmental implications, including potential disruption to existing ecosystems and landscapes.

Future Prospects and Innovations

The future of wireless and dynamic charging holds promising potential for transforming the EV charging landscape:

1. Integration with Renewable Energy:

- Solar and Wind Power: Wireless charging systems integrated with renewable energy sources, such as solar and wind power, can enhance sustainability and reduce reliance on fossil fuels.

- Smart Grid Integration: Combining wireless charging with smart grid technology can optimize energy distribution, manage demand, and reduce peak loads.

2. Advancements in Materials and Technology:

- High-Efficiency Materials: Research into advanced materials, such as high-temperature superconductors, could improve the efficiency and power output of wireless charging systems.

- Adaptive Control Systems: Developing adaptive control systems that can dynamically adjust power levels and optimize energy transfer based on real-time conditions will enhance the effectiveness of wireless charging.

3. Pilot Programs and Real-World Applications:

- Urban Deployment: Pilot programs in urban environments can demonstrate the feasibility and benefits of wireless and dynamic charging, paving the way for broader adoption.

- Public-Private Partnerships: Collaboration between governments, industry stakeholders, and research institutions can accelerate the development and deployment of wireless charging infrastructure.

In conclusion, wireless and dynamic charging represent the cutting edge of EV charging technology, offering unprecedented convenience and the potential to revolutionize how we power our electric vehicles. By addressing the challenges and leveraging ongoing innovations, we can create a future where seamless, cable-free charging becomes the norm, supporting the widespread adoption of EVs and contributing to a cleaner, more sustainable transportation system.

CHAPTER 12

SMART CHARGING SOLUTIONS

Introduction to Smart Charging Solutions

As electric vehicles (EVs) become more prevalent, the integration of smart charging solutions is essential to maximize the benefits of EV adoption and ensure the stability and efficiency of the electrical grid. Smart charging solutions encompass advanced technologies and strategies that optimize energy use, reduce grid impact, and support renewable energy integration. Two key components of smart charging solutions are grid integration and vehicle-to-grid (V2G) technology, both of which play a pivotal role in creating a sustainable and resilient energy ecosystem.

Grid Integration: Optimizing Energy Use and Reducing Grid Impact

Grid integration refers to the seamless incorporation of EV charging infrastructure with the electrical grid, allowing for efficient management of energy resources. Enhanced grid integration can optimize energy use, reduce peak demand, and improve overall grid stability. Here's how it works:

1. Smart Grids and Their Components:

- Advanced Metering Infrastructure (AMI): AMI involves smart meters that provide real-time data on energy consumption and allow for two-way communication between utilities and consumers. This data is crucial for managing and optimizing EV charging.

- Demand Response (DR) Programs: DR programs incentivize consumers to reduce or shift their energy use during peak demand periods. For EV owners, this can mean charging their vehicles during off-peak hours when electricity is cheaper and more abundant.

- Distributed Energy Resources (DERs): DERs, such as solar panels and home batteries, can be integrated with EV charging systems to provide additional power sources and reduce reliance on the grid.

2. Benefits of Grid Integration:

- Load Management: Smart charging systems can manage the load on the grid by controlling the timing and rate of EV charging. This helps prevent grid overload and reduces the need for expensive grid upgrades.

- Cost Savings: By shifting EV charging to off-peak hours, consumers can take advantage of lower electricity rates, resulting in significant cost savings.

- Enhanced Reliability: Smart grid technology enhances the reliability and resilience of the grid by providing real-time data and enabling quick response to fluctuations in energy demand.

3. Strategies for Effective Grid Integration:

- Time-of-Use (TOU) Pricing: TOU pricing structures encourage EV owners to charge their vehicles during off-peak hours when electricity is cheaper. This not only saves money for consumers but also helps balance the load on the grid.

- Smart Charging Stations: Smart charging stations can communicate with the grid and adjust charging rates based on grid conditions. These stations can prioritize charging during periods of low demand and reduce or delay charging during peak demand.

- Energy Management Systems (EMS): EMS can integrate various energy sources, including renewable energy and DERs, to optimize energy use and minimize grid impact. These systems can manage the charging of multiple EVs, ensuring efficient and balanced energy distribution.

V2G Technology: Enhancing Grid Stability and Supporting Renewable Energy Integration

Vehicle-to-grid (V2G) technology enables bidirectional energy flow between EVs and the grid, allowing EVs to supply power back to the grid when needed. This capability enhances grid stability, supports renewable energy integration, and provides additional value to EV owners.

1. How V2G Technology Works:

- Bidirectional Chargers: V2G systems require bidirectional chargers that can both charge the EV battery and discharge energy back to the grid. These chargers facilitate the two-way flow of electricity.

- Communication Systems: V2G technology relies on advanced communication systems that allow the EV and the grid to exchange information. This includes data on grid conditions, energy demand, and battery status.

- Energy Management Platforms: These platforms coordinate the energy flow between the EV and the grid, ensuring that energy is transferred efficiently and in a way that benefits both the grid and the EV owner.

2. Benefits of V2G Technology:

- Grid Stability: By supplying energy back to the grid during periods of high demand, V2G technology helps stabilize the grid and prevent outages. EVs can act as mobile energy storage units, providing a valuable resource for grid operators.

- Renewable Energy Integration: V2G technology supports the integration of renewable energy sources, such as solar and wind power, by providing a flexible and responsive energy storage solution. EVs can store excess renewable energy during periods of low demand and supply it back to the grid when needed.

- Economic Incentives: EV owners can earn financial incentives by participating in V2G programs. These incentives can offset the cost of EV ownership and provide additional income for vehicle owners.

3. Challenges and Considerations for V2G Implementation:

- Infrastructure Requirements: Implementing V2G technology requires significant investment in infrastructure, including bidirectional chargers and advanced communication systems. Coordination between utilities, EV manufacturers, and charging network operators is essential.

- Battery Degradation: Frequent cycling of the battery for V2G purposes can potentially impact battery lifespan. Research and development are ongoing to mitigate these effects and ensure that V2G technology is both efficient and durable.

- Regulatory and Market Barriers: Regulatory frameworks and market structures need to be adapted to accommodate V2G technology. This includes developing standards for bidirectional charging and creating market mechanisms that value and reward the services provided by V2G.

Future Prospects of Smart Charging Solutions

The future of smart charging solutions is marked by continuous innovation and the development of new technologies that further enhance the benefits of grid integration and V2G:

1. Integration with Smart Cities:

- Urban Planning: Smart charging solutions will play a crucial role in the development of smart cities, where integrated energy systems optimize resource use and enhance the quality of life for residents.

- Public Charging Infrastructure: Expansion of public charging infrastructure, equipped with smart charging capabilities, will support the growing number of EVs in urban environments.

2. Advancements in Battery Technology:

- Solid-State Batteries: The development of solid-state batteries, with higher energy density and longer lifespans, will enhance the viability and efficiency of V2G technology.

- Second-Life Batteries: Repurposing used EV batteries for grid storage can provide additional value and support renewable energy integration.

3. Blockchain and Artificial Intelligence:

- Blockchain Technology: Blockchain can facilitate secure and transparent transactions in V2G systems, enabling efficient energy trading and data management.

- Artificial Intelligence (AI): AI can optimize smart charging systems by predicting energy demand, managing grid resources, and ensuring efficient energy distribution.

4. Policy and Regulatory Support:

- Incentive Programs: Government incentives and subsidies can accelerate the adoption of smart charging solutions and V2G technology.

- Standardization: Developing standardized protocols for bidirectional charging and smart grid communication will facilitate interoperability and streamline implementation.

In conclusion, smart charging solutions, including enhanced grid integration and V2G technology, are essential for maximizing the benefits of EV adoption and creating a sustainable and resilient energy ecosystem. By leveraging these advanced technologies, we can optimize energy use, reduce grid impact, support renewable energy integration, and provide additional value to EV owners. The ongoing innovation and collaboration among stakeholders will pave the way for a future where smart charging solutions are an integral part of our energy infrastructure, contributing to a cleaner, more sustainable world.

REFERENCES

National Electrical Code (NEC)

- NFPA 70: National Electrical Code (NEC): This is a regionally adoptable standard for the safe installation of electrical wiring and equipment in the United States. It sets the foundation for electrical safety in residential, commercial, and industrial occupancies.

- Source: [National Fire Protection Association (NFPA)](https://www.nfpa.org/NEC)

Local Building Codes and Regulations

- International Building Code (IBC): While the NEC provides electrical safety standards, the IBC addresses overall building safety, including structural, fire, and accessibility standards.

- Source: [International Code Council (ICC)](https://www.iccsafe.org/)

- Local Building Codes: Each municipality or state may have additional or more stringent requirements that go beyond national standards. It's essential to check with local building departments or regulatory bodies for specific requirements.

- Example: Local amendments to the IBC and NEC that might be specific to regions like California (California Building Standards Code), New York City Building Code, etc.

Manufacturer Installation Manuals and Guidelines

- Tesla Supercharger Installation Manual: This guide provides detailed instructions for the installation of Tesla's Supercharger stations, covering site selection, electrical requirements, and installation procedures.

- Source: Tesla official website or through direct contact with Tesla's installation support.

- ChargePoint DC Fast Charger Installation Guide: This manual includes comprehensive steps for installing ChargePoint's DC fast chargers, including site preparation, electrical connections, and commissioning.

- Source: [ChargePoint Installation Resources](https://www.chargepoint.com)

- ABB Terra DC Fast Charger Installation Manual: ABB provides detailed installation manuals for their Terra series DC fast chargers, which cover electrical and mechanical installation, commissioning, and maintenance.

- Source: [ABB EV Charging Solutions](https://new.abb.com/ev-charging)

- EVBox Troniq Modular Installation Guide: This guide offers a step-by-step process for installing EVBox's modular DC fast charging stations, including hardware setup, software configuration, and safety checks.

- Source: [EVBox Installation Support](https://www.evbox.com)

- Siemens VersiCharge Installation Manual: Siemens provides installation guidelines for their VersiCharge line of DC fast chargers, ensuring compliance with safety and operational standards.

- Source: [Siemens EV Charging Solutions](https://www.siemens.com)

In conclusion, adhering to national and local electrical and building codes, as well as manufacturer-specific installation guidelines, is essential for the safe and effective deployment of DC fast chargers. These references provide the necessary frameworks and detailed instructions to ensure compliance and optimal performance of EV charging infrastructure.

INDEX

By using this index, readers can quickly locate key topics and terms covered in the "DC Fast Chargers Installation Guide." This comprehensive reference section ensures that users can find the information they need efficiently.

9 798348 125400